he ESSENTIALS® of

MODERN ALGEBRA

D0557865

Lutfi A. Lutfiyya, Ph.D.
Associate Professor of Mathematics
Kearney State College, Nebraska

Research and Education Association
61 Ethel Road West
Piscataway, New Jersey 08854

THE ESSENTIALS OF
MODERN ALGEBRA ®

Printed in the United States of America

Library of Congress Catalog Card Number 89-61501

International Standard Book Number 0-87891-681-4

ESSENTIALS is a registered trademark of
Research and Education Association, Piscataway, New Jersey 08854

WHAT "THE ESSENTIALS" WILL DO FOR YOU

This book is a review and study guide. It is comprehensive and it is concise.

It helps in preparing for exams, in doing homework, and remains a handy reference source at all times.

It condenses the vast amount of detail characteristic of the subject matter and summarizes the **essentials** of the field.

It will thus save hours of study and preparation time.

The book provides quick access to the important facts, principles, theorems, concepts, and equations in the field.

Materials needed for exams can be reviewed in summary form – eliminating the need to read and re-read many pages of textbook and class notes. The summaries will even tend to bring detail to mind that had been previously read or noted.

This "ESSENTIALS" book has been prepared by an expert in the field, and has been carefully reviewed to assure accuracy and maximum usefulness.

Dr. Max Fogiel
Program Director

CONTENTS

CHAPTER 1

SET THEORY

1.1 INTRODUCTION

If S is a collection of objects, then the objects are called the elements of S. We write

$$x \in S$$

to mean x is an element of S, and we write

$$x \notin S$$

to mean x is not an element of S.

We may specify a set by stating in words what its elements are. Another way of specifying a set is to exhibit its elements, usually enclosed in braces. Thus, $\{x\}$ indicates the set consisting of the single element x; $\{x, y\}$ indicates the set consisting of the two elements, x and y; and if P is the set of all positive integers, by writing

$$k = \{a \mid a \in P, a \text{ divisible by } 2\}$$

we mean that k consists of all elements, a, having the properties indicated after the vertical bar. Thus,

$$k = \{2, 4, 6, 8, \dots\}.$$

1.2   EQUALITY OF SETS

A set is specified by its elements. Thus, two sets A and B are said to be equal if and only if they have the same elements, and we write

$$A = B.$$

1.3   THE EMPTY SET

The need arises for a very peculiar set, namely the set which has no elements at all. This set is called the null or empty set. This set is denoted by the symbol ϕ (Phi). For example, the set consisting of all college students in the USA who are less than 8 years old.

1.4   SUBSETS

Consider two sets, S and T. If every element of S is also an element of T, then S is called a subset of T, and we write

$$S \subseteq T \quad \text{or} \quad T \supseteq S$$

The empty set, ϕ, has the property that it is a subset of every set S. Also, $S \subseteq S$ for every set S.

A finite set, S, with n elements has 2^n subsets.

1.5   PROPER SUBSETS

If S and T are sets such that $S \subseteq T$, and $S \neq T$, the S is called a proper subset of T. In this case, we write,

$$S \subset T$$

to denote that S is a proper subset of T. Hence, if

$$S \subseteq T, \quad \text{and} \quad T \subseteq S$$

then,

$$S = T.$$

1.6   OPERATIONS ON SETS

Let S and T be sets, then

a)   the union of S and T is the set $S \cup T$ given by

$$S \cup T = \{x \mid x \in S, \text{ or } x \in T\}$$

b)   the intersection of S and T is the set $S \cap T$ given by

$$S \cap T = \{x \mid x \in S \text{ and } x \in T\}$$

c)   the complement of T in S, or the difference between S and T is the set $S - T$ given by

3

$$S - T = \{x \mid x \in S, \text{but } x \notin T\}$$

In general, $S - T \neq T - S$.

1.7   VENN DIAGRAMS

Sets can be represented pictorially by what are called Venn Diagrams. The above sets in section 1.3 can be represented as follows:

a) $S \cup T$

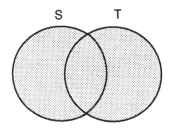

is the shaded area.

b) $S \cap T$

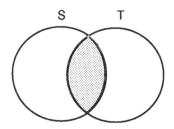

is the shaded area.

c) $S - T$

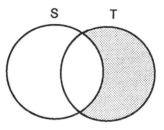

is the shaded area.

$T - S$

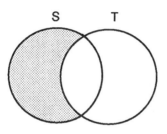

is the shaded area.

1.8   POWER SETS

Let S be any set. The power set of S, denoted by $*P(S)$, is the set of all subsets of S and is written as

$$P(S) = \{A \mid A \subseteq S\}$$

If S is a finite set having n elements, then $P(S)$ has 2^n elements. For example, if

$$S = \{a, b, c\}$$

then

$$P(S) = \{ \{a\}, \{b\}, \{c\}, \{a,b\}, \{a,c\}, \{b,c\}, \{a,b,c\}, \phi \}.$$

1.9   PARTITIONS OF SETS

Let S be a nonempty set. A collection, P, of nonempty subsets of S is called a partition of S if

a) S is the union of all the sets in P. That is,

$$\bigcup_{A \in P} A = S$$

b)   for any $A, B \in P$, either $A = B$, or $A \cap B = \phi$.

If $S = \{a, b, c, d, e, f\}$, and

$$A_1 = \{a, b\}, A_2 = \{c\}, A_3 = \{d, e, f\}$$

then, $P = \{A_1, A_2, A_3\}$ is a partition of S since,

a) $A_1 \cup A_2 \cup A_3 = S$

b) $A_1 \cap A_2 = \phi, A_1 \cap A_3 = \phi$, and $A_2 \cap A_3 = \phi$.

1.10   THE CARTESIAN PRODUCT OF SETS

Let S and T be nonempty sets. The Cartesian product of S and T is the set $S \times T$ given by

$$S \times T = \{(a, b) \mid a \in S \text{ and } b \in T\}.$$

In general, $S \times T \neq T \times S$.

Note: Two ordered pairs (x_1, y_1) and (x_2, y_2) are equal, i.e. $(x_1, y_1) = (x_2, y_2)$ if and only if $x_1 = x_2$ and $y_1 = y_2$.

CHAPTER 2

MAPPINGS, OPERATIONS, AND RELATIONS

2.1   MAPPINGS

Let S and T be nonempty sets. A mapping, α, from S to T is a subset of $S \times T$ such that

a)   for each $x \in S$, there exists a unique element $y \in T$, such that $(x, y) \in \alpha$.

b)   If (x, y) and $(x, z) \in \alpha$, then $y = z$.

The set S is called the domain of α, and the set T is called the codomain of α, sometimes the range of α.

If α is a mapping from S to T, we write

$$\alpha : S \to T$$

and if $x \in S$, then $\alpha(x)$ is called the image of x in T under α.

2.2 SURJECTIVE, INJECTIVE, AND BIJECTIVE MAPPINGS

Consider the mapping

$$\alpha : S \to T$$

a) α is said to be surjective, or onto, if every element $y \in T$ is the image under α of some element $x \in S$. That is, given $y \in T$, there exists $x \in S$, such that $y = f(x)$.

b) α is called injective, or one-to-one, if different elements of S always have different images in T. That is, if x_1 and $x_2 \in S$, such that $x_1 \neq x_2$, then $f(x_1) \neq f(x_2)$.

c) α is called a bijection, or one-to-one and onto, or a one-to-one correspondence, if it is both injective and surjective.

2.3 MAPPING COMPOSITION

Let

$$\alpha : S \to T, \text{ and}$$

$$\beta : T \to U$$

be mappings. The composition mapping, $\beta \circ \alpha$, is the mapping from S to U defined by

$$(\beta \circ \alpha)(x) = \beta(\alpha(x))$$

for every $x \in S$.

The composition mapping, $\beta \circ \alpha$, is diagrammed as follows:

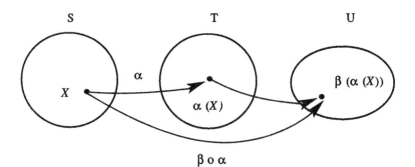

Note that the domain of β must contain the range of α before the composition mapping, $\beta \circ \alpha$, is defined.

2.4 SOME PROPERTIES OF MAPPING COMPOSITION

Assume that

$$\alpha : S \rightarrow T, \text{ and}$$
$$\beta : T \rightarrow U$$

are mappings.

a) If α and β are both surjective, then $\beta \circ \alpha$ is surjective.

b) If $\beta \circ \alpha$ is surjective, then β is surjective.

c) If α and β are both injective, then $\beta \circ \alpha$ is injective.

d) If $\beta \circ \alpha$ is injective, then α is injective.

2.5   IDENTITY MAPPINGS

If S is any nonempty set, then the mapping

$$I_s : S \to S$$

defined by $I_s(x) = x$ for every $x \in S$ is called the identity mapping on S.

The identity mapping is surjective and injective.

2.6   INVERTIBLE MAPPINGS

Consider a mapping,

$$\alpha : S \to T$$

A mapping $\beta : T \to S$ is an inverse of α if both $\beta \circ \alpha = I_s$ and $\alpha \circ \beta = I_T$. A mapping is said to be invertible if it has an inverse. If a mapping is invertible, then its inverse is unique.

Note that a mapping is invertible if and only if it is injective and surjective.

2.7   BINARY OPERATIONS

Let S be a given nonempty set. A binary operation, or simply an operation, $*$, on S is a mapping from $S \times S$ to S.

If $*$ is an operation on a set S, then it is always assumed that $x * y \in S$ for all $x, y \in S$.

2.8 SOME PROPERTIES OF OPERATIONS

Let $*$ be a binary operation defined on the set S. Then,

(i)   the operation $*$ is said to be commutative if and only if

$$x * y = y * x$$

for all $x, y \in S$.

(ii)   the operation $*$ is said to be an associative operation if and only if

$$(x * y) * z = x * (y * z)$$

for all $x, y, z \in S$.

(iii)   an element $e \in S$ is said to be an identity for the operation $*$ if and only if

$$x * e = e * x = x$$

for every $x \in S$.

(iv)   if e is the identity of the operation $*$, and $a \in S$, then the element $b \in S$ is an inverse of the element a relative to $*$ if and only if

$$a * b = b * a = e.$$

2.9 RELATIONS

Let S be a given nonempty set. We then say that a relation, R, is defined on S if for each ordered pair (x, y) of elements of S, it is true or false that x is in the relation R to y. It is customary to write

$$x R y$$

to indicate that x is in the relation R to y.

2.10 EQUIVALENCE RELATIONS

Let S be a nonempty set. An equivalence relation on S is a subset R of $S \times S$ satisfying the following three conditions:

(i) $(a, a) \in R$, for all $a \in S$.

(ii)   If $(a, b) \in R$, then $(b, a) \in R$.

(iii)   If $(a, b) \in R$, and $(b, c) \in R$, then $(a, c) \in R$.

Condition (i) is expressed by saying that R is reflexive, (ii) by saying that R is symmetric, and (iii) by saying that R is transitive.

2.11 EQUIVALENCE CLASSES

If R is an equivalence relation on a set S, then the equivalence class of an element $x \in S$, denoted by $[x]$, is defined as

$$[x] = \{y \mid y \in S \text{ and } y R x\}.$$

If R is an equivalence relation on a nonempty set S, then

a) $S = \bigcup_{x \in S} [x]$

b) Two equivalence classes, $[x]$ and $[y]$, are either equal or disjoint. That is, either $[x] = [y]$ or $[x] \cap [y] = \phi$.

CHAPTER 3

BASIC PROPERTIES OF THE INTEGERS

3.1   PEANO'S POSTULATES

During the 19th century, the mathematician G. Peano formulated a set of postulates for the positive integers. They are known as Peano's Postulates, and may be stated as follows:

1   There is a positive integer 1.

2.   Each positive integer n has a successor n'.

3.   1 is not the successor of any positive integer.

4.   If $m' = n'$, then $m = n$.

5.   **The Induction Postulate**–If S is a set of positive integers, then S contains all of the positive integers if it has the following properties:

         a)   1 is in S

         b) $n \in S$ always implies $n' \in S$.

3.2 BASIC AXIOMS OR POSTULATES

It is possible to use Peano's postulates as a basis from which to develop all of the familiar properties of the positive integers. The set of all integers may then be constructed from the set of all positive integers. However, we assume that there is a set of Z of elements, called the integers, that satisfies the following conditions:

1. ADDITION POSTULATES

There is a binary operation defined in Z that is called addition and denoted by +, and has the following properties:

a) Z is closed under addition.

b)   Addition is associative.

c) Z contains an element 0 that is an identity element for addition.

d)   For each $x \in Z$, there is an additive inverse of x in Z, denoted by $-x$, such that $x + (-x) = 0 = (-x) + x$.

e)   Addition is commutative.

2. MULTIPLICATION POSTULATES

There is a binary operation defined in Z that is called multiplication and denoted by $\cdot$, and has the following properties:

a) Z is closed under multiplication.

b) Multiplication is associative.

c) Z contains an element 1 that is different from 0 and that is an identity for multiplication.

d) Multiplication is commutative.

3. THE DISTRIBUTIVE LAW

For all elements $x, y, z \in Z, x \cdot (y + z) = x \cdot y + x \cdot z$.

4. Z contains a subset Z^+, called the set of positive integers, that has the following properties:

a) Z^+ is closed under addition.

b) Z^+ is closed under multiplication.

c) For each $x \in Z^+$, one and only one of the following statements is true:

   (i) $x \in Z^+$

   (ii) $x = 0$

   (iii) $-x \in Z^+$

5. INDUCTION POSTULATE

Let S be a set of positive integers with the following properties:

a) $1 \in S$

b)   If k is any positive integer such that $k \in S$, then also $k + 1 \in S$

Then S consists of the set of all positive integers.

3.3   PRINCIPLE OF MATHEMATICAL INDUCTION

Suppose that there is associated with each positive integer n a statement S_n such that the following hold:

(i) S_1 is true

(ii)   If k is any positive integer such that S_k is true, then S_{k+1} is true.

Then S_n is true for all positive integers n.

Thus, a proof that a statement S_n is true for all positive integers n consists of three steps:

1)   The statement is proved or verified true for $n = 1$.

2)   For a positive integer k, the statement is assumed true for all positive integers $m < k$.

3)   Under this assumption (Step 2), the statement is proved to be true for $m = k$.

3.4   DIVISIBILITY

Given two integers m and n, with $m \neq 0$, we say that m divides n, written symbolically as m/n, if there exists an integer c such that $n = mc$.

There are alternate ways to express this, namely, m is a divisor of n, m is a factor of n, n is a multiple of m, n is divisible by m.

If no such c exists, we say that m does not divide n (or: m is not a divisor of n, m is not a factor of n, n is not a multiple of m, n is not divisible by m), and write $m \nmid n$.

The basic elementary properties of divisibility may be stated as follows:

a) $1 \mid n$ for all integers n.

b)   If m is an integer such that $m \neq 0$, then $m \mid 0$.

c)   If $m \mid n$, and $m \mid q$, then $m \mid (un + vq)$ for all integers m, n, q, u, and v.

d)   If $m \mid n$ and $n \mid q$, then $m \mid q$.

e)   If $m \mid 1$, then $m = 1$ or $m = -1$.

f)   If $m \mid n$, and $n \mid m$, then $m = \mp n$.

3.5   PRIME FACTORS AND THE GREATEST COMMON DIVISOR

Given two integers m and n, not both 0, then an integer d is called the greatest common divisor of m and n if the following conditions are satisfied:

1) d is a positive integer

2) $d \mid m$ and $d \mid n$

3)   If c is any integer such that $c \mid m$ and $c \mid n$, then $c \mid d$

We write this d as $d = (m, n)$.

Thus, for instance, $(24, 9) = 3 = (-24, 9) = (24, -9) = (-24, -9)$, and $(24, 7) = 1$.

If the greatest common divisor of two integers m and n is 1, i.e. $(m, n) = 1$, we say that m and n are relatively prime.

An integer $p > 1$ is said to be prime when 1 and p are its only positive divisors. If $p > 1$ is not prime, we say p is composite. Thus, to say that $n > 1$ is composite means that we can write $n = ab$ with $1 < a < n$ and $1 < b < n$, while saying that $n > 1$ is prime means that n cannot be written in this form.

3.6   THE LEAST COMMON MULTIPLE

If a and b are non-zero integers, then there is a unique positive integer m such that

a) $a \mid m$ and $b \mid m$

b)   If c is an integer such that $a \mid c$ and $b \mid c$, then $m \mid c$

20

Property (a) states that m is a common multiple of a and b; property (b) ensures that m is the least positive such multiple. Therefore, the integer m is called the least common multiple of a and b. It is denoted by $[a, b]$. Examples are $[4, -6] = 12$; $[-7, 7] = 7$; and $[25, 33] = 825$.

3.7 THE DIVISION ALGORITHM

If m and n are integers with $n > 0$, then there exist unique integers q and r such that

$$m = qn + r , \ 0 \le r < n.$$

Thus, for instance, if $m = 53$ and $n = 6$, then

$$53 = 6 \cdot 8 + 5.$$

In the Division Algorithm, the integer q is called the quotient and r is called the remainder in the division of m by n.

3.8 EUCLID'S ALGORITHM

Euclid's Algorithm provides us with a systematic procedure for computing the greatest common divisor of any two integers m and n.

By the Division Algorithm, if $n > 0$, there exist unique integers q_1 and r_1 such that

$$m = q_1 n + r_1, 0 \le r_1 < n.$$

If $r_1 = 0$, then $n \mid m$ and $(m, n) = n$. If $r_1 \neq 0$, we can apply the Division Algorithm again, getting integers q_2 and r_2 such that

$$n = r_1 q_2 + r_2, \, 0 \leq r_2 < r_1.$$

Repeated application of the Division Algorithm in this way produces a sequence of pairs of integers $q_1, r_1; q_2, r_2; q_3, r_3; \dots$ such that

$$m = nq_1 + r_1 \, , \, 0 \leq r_1 < n$$
$$n = r_1 q_2 + r_2 \, , \, 0 \leq r_2 < r_1$$
$$r_1 = r_2 q_3 + r_3 \, , \, 0 \leq r_3 < r_2$$

$$\dots \dots$$

Because each remainder is non-negative, and

$$r_1 > r_2 > r_3 > \dots,$$

we must eventually reach a remainder of zero. If r_{k+1} denotes the first zero remainder, then the process terminates with

$$r_{k-2} = r_{k-1} q_k + r_k \, , \, 0 \leq r_k < r_{k-1}$$
$$r_{k-1} = r_k q_{k+1}$$

and r_k, the last non-zero remainder, is the greatest common divisor of m and n. That is, $(m, n) = r_k$.

Here is Euclid's Algorithm applied to compute $(1001, 357)$:

$$1001 = 357 \cdot 2 + 287$$
$$357 = 287 \cdot 1 + 70$$

$$287 = 70 \cdot 4 + 7$$
$$70 = 10 \cdot 7$$

Therefore, $(1001, 357) = 7$.

3.9 THE FUNDAMENTAL THEOREM OF ARITHMETIC

Every positive integer n is either 1 or can be expressed as a product of prime integers, and this factorization is unique except for the order of the factors.

This theorem, sometimes called the unique factorization theorem, can be used to describe a standard form for a positive integer n. If $P_1, P_2 \ldots, P_r$ are all distinct prime factors of n, arranged in order of magnitude, so that

$$P_1 < P_2 < P_3 < \ldots < P_r,$$

then all repeated factors may be collected together and expressed by use of exponents to yield

$$n = P_1^{m_1} P_2^{m_2} P_3^{m_3} \ldots P_r^{m_i}$$

where each m_i is a positive integer. Each m_i is called the multiplicity of P_i, and this factorization is known as the standard form for n. Thus, for instance, $504 = 2^3 \cdot 3^2 \cdot 7$.

3.10 CONGRUENCE OF INTEGERS

Let n be a positive integer such that $n > 1$. For integers x and

y, we say x is congruent to y modulo n if and only if $x - y$ is a multiple of n. We write

$$x \equiv y \ (\text{mod } n)$$

to indicate that x is congruent to y modulo n.

Thus, $x \equiv y \ (\text{mod } n)$ if and only if n divides $x - y$, and this is equivalent to $x - y = nq$ or $x = y + nq$ for some integer q.

Another way to describe this relation is to say that x and y yield the same remainder when each is divided by n. In particular, each integer is congruent to its remainder when divided by n. This means that any integer x is congruent to one of the integers:

$$0, 1, 2, 3, \ldots, n - 1$$

when x is divided by n.

Congruence modulo n is an equivalence relation on the set of all integers, Z. As with any equivalence relation, the equivalence classes for congruence modulo n form a partition of Z; that is, they separate Z into mutually disjoint subsets. These subsets are called congruence classes, or residue classes. Referring to our discussion above concerning remainders, we see that there are n distinct congruence classes modulo n, given by

$$[0] = \{\ldots, -2n, -n, 0, n, 2n, \ldots\}$$
$$[1] = \{\ldots, -2n + 1, -n + 1, 1, n + 1, 2n + 1, \ldots\}$$
$$[2] = \{\ldots, -2n + 2, -n + 2, 2, n + 2, 2n + 2, \ldots\}$$
$$\vdots$$
$$[n-1] = \{\ldots, -n - 1, -1, n - 1, 2n - 1, 3n - 1, \ldots\}$$

When $n = 4$, these classes appear as:

$$[0] = \{ \ldots, -8, -4, 0, 4, 8, \ldots \}$$
$$[1] = \{ \ldots, -7, -3, 1, 5, 9, \ldots \}$$
$$[2] = \{ \ldots, -6, -2, 2, 6, 10, \ldots\}$$
$$[3] = \{ \ldots, -5, -1, 3, 7, 11, \ldots\}$$

3.11   CONGRUENCE CLASSES

In connection with the relation congruence modulo n, we have observed that there are n distinct congruence classes. Let Z_n denote this set of congruence classes:

$$Z_n = \{ \ [0], [1], [3], \ldots, [n-1] \ \}$$

When addition, $\oplus$, and multiplication, $\odot$, are defined in a natural and appropriate manner in Z_n, these sets provide us with very useful examples for work with the theory of groups.

1.   Addition, $\oplus$: Consider the rule given by

$$[a] \oplus [b] = [a + b]$$

a)   This rule defines an addition which is a binary operation on Z_n.

b)   Addition is associative in Z_n,

$$([a] \oplus [b]) \oplus [c] = [a] \oplus ([b] \oplus [c])$$

for all $[a], [b], [c] \in Z_n$.

c) Z_n has the additive identity [0].

d)   Each $[a] \in Z_n$ has an additive inverse $[-a] \in Z_n$.

e)   Addition is commutative in Z_n,

$$[a] \oplus [b] = [b] \oplus [a]$$

for all $[a]$, $[b] \in Z_n$.

The complete addition table for Z_4 is as follows:

$\oplus$	[0]	[1]	[2]	[3]
[0]	[0]	[1]	[2]	[3]
[1]	[1]	[2]	[3]	[0]
[2]	[2]	[3]	[0]	[1]
[3]	[3]	[0]	[1]	[2]

2. Multiplication, $\odot$: Consider the rule for multiplication in Z_n given by

$$[a] \odot [b] = [ab]$$

a)   Multiplication as defined by this rule is a binary operation on Z_n.

b)   Multiplication is associative in Z_n:

$$([a] \odot [b]) \odot [c] = [a] \odot ([b] \odot [c])$$

for all $[a]$, $[b]$, $[c] \in Z_n$.

c) Z_n has a multiplicative identity $[1] \in Z_n$.

d)   Multiplication is commutative in Z_n:

$$[a] \odot [b] = [b] \odot [a]$$

for all $[a]$, $[b] \in Z_n$.

The complete multiplication table for Z_4 is as follows:

$\odot$	$[0]$	$[1]$	$[2]$	$[3]$
$[0]$	$[0]$	$[0]$	$[0]$	$[0]$
$[1]$	$[0]$	$[1]$	$[2]$	$[3]$
$[2]$	$[0]$	$[2]$	$[0]$	$[2]$
$[3]$	$[0]$	$[3]$	$[2]$	$[1]$

The third row of the table shows that $[2]$ is a non-zero element of Z_4 that has no multiplicative inverse; there is no $[x] \in Z_4$ such that $[2] \odot [x] = [1]$. Another interesting point in connection with this table is that the equality $[2] \odot [2] = [0]$ shows that in Z_n, the product of non-zero factors may be zero.

CHAPTER 4

GROUP THEORY

4.1 DEFINITION OF A GROUP

A non-empty set G on which there is defined a binary operation $*$ is called a group with respect to this operation provided the following properties hold:

(1) G is closed under $*$: if $a, b \in G$, then $a * b \in G$.

(2) $*$ is associative in G: for all $a, b, c \in G$, $(a * b) * c = a * (b * c)$.

(3) G has an identity element: there exists a special element $e \in G$ such that

$$e * a = a * e = a$$

for all $a \in G$. (e is also called the unit element of G).

(4)   Existence of inverses: for every element $a \in G$, there exists an element $b \in G$ such that

$$a * b = b * a = e.$$

28

This element b is written as a^{-1} and is called the inverse of a.

4.1.1   ORDER OF A GROUP

If a group G has a finite number of elements, then G is called a finite group, or a group of finite order. The number of elements in G is called the order of G and is denoted by $0(G)$. If G does not have a finite number of elements, then G is called an infinite group, or a group of infinite order.

4.2   PROPERTIES OF GROUPS

Let G be an arbitrary group, with respect to an operation $*$; then the following properties hold in G:

(i)   The identity element, e, of G is unique.

(ii)   If $a \in G$, then a has a unique inverse in G.

(iii)   If $a, b, c \in G$ such that $a * b = a * c$, then $b = c$.

(iv)   If $a, b, c \in G$ such that $b * a = c * a$, then $b = c$.

(v)   If $a, b \in G$, then there exists a unique element $x \in G$ such that $a * x = b$, and a unique element $y \in G$ such that $y * a = b$. In fact, $x = a^{-1} * b$ and $y = b * a^{-1}$.

(vi)   If $a \in G$, then $(a^{-1})^{-1} = e$.

(vii)   If $a, b \in G$, then $(a * b)^{-1} = b^{-1} * a^{-1}$.

4.2.1 INTEGRAL EXPONENTS OF ELEMENTS OF A GROUP

Let G be a group with the binary operation written as multiplication. For any $a \in G$, and integers n and m, non-negative integral exponents are defined by

$$a^0 = e, \quad a^{-n} = (a^{-1})^n, \quad a * a^n = a^{n+1},$$
$$a^m * a^n = a^{m+n}, (a^m)^n = a^{mn}.$$

4.2.2 MULTIPLES OF ELEMENTS OF A GROUP

Let G be a group with the binary operation written as addition. For elements $a, b \in G$ and integers m and n, multiples of a and b are given by

$$0a = 0, \quad 1a = a, \quad (m + 1)a = ma + a,$$
$$(-m)a = m(-a), \quad ma + na = (m + n)a,$$
$$n(ma) = (nm)a, \quad n(a + b) = na + nb$$

The notation ma here does not represent a product of m and a, but rather a sum

$$ma = a + a + a + \ldots + a.$$

4.2.3 GENERALIZED ASSOCIATIVE LAW

Let $n \geq 2$ be a positive integer, and let $a_1, a_2, a_3, \ldots, a_n$ be elements of a group G, then for any positive integer m such that $1 \leq m < n$,

$$(a_1 a_2 \ldots a_m)(a_{m+1} a_{m+2} \ldots a_n) = a_1 a_2 a_3 \ldots a_n.$$

4.3   EXAMPLES OF GROUPS

1.   Let Z be the set of all integers, and let $*$ be the ordinary addition, $+$, in Z. Then Z is a group with respect to $+$.

2.   Let G be the set $\{e, a, b, c\}$ with an operation of multiplication defined by the following table:

·	e	a	b	c
e	e	a	b	c
a	a	e	c	b
b	b	c	e	a
c	c	b	a	e

Then G is a group with respect to the defined operation.

3.   Recall that $Z_4 = \{[0], [1], [2], [3]\}$ and the operation $\oplus$, addition of the equivalence classes. Z_4 is a group with respect to $\oplus$.

4.   Recall $Z_5 - \{[0]\} = \{[1], [2], [3], [4]\}$ and the operation $\odot$, multiplication of the equivalence classes. $Z_5 - \{[0]\}$ is a group with respect to $\odot$.

5.   Let S be any non-empty set, and let $A(S)$ be the set of all invertible mappings from S to S. $A(S)$ is a group with respect to the composition of mappings, 0, as the operation.

6.   If $G = \{1, -1, i, -i\}$ where $i = \sqrt{-1}$, and the operation "$\cdot$" is ordinary multiplication, then G is a group with respect to "$\cdot$".

4.4   ABELIAN GROUPS

If in a group G with operation $*$,

$$a * b = b * a$$

for all $a, b \in G$, then G is said to be an abelian group.

A group that is not abelian is called a non-abelian or a non-commutative group.

4.4.1   SOME PROPERTIES OF ABELIAN GROUPS

Let G be an abelian group with respect to the operation $*$. Then for all $a, b \in G$ and any integer n,

(i) $(a * b)^{-1} = a^{-1} * b^{-1}$

(ii) $(a * b)^n = a^n * b^n$

(iii)   For any group G, if $0(G) < 6$, then G is an abelian group.

4.5   SUBGROUPS

Let G be a group with respect to an operation $*$. Then a non-empty subset H of G is called a subgroup of G if H forms a group with respect to the operation $*$.

4.5.1 TRIVIAL AND NON-TRIVIAL SUBGROUPS

Let G be a group with respect to an operation $*$ and e be the identity element of G. Then the subsets $H = \{e\}$ and $H = G$ are always subgroups of G. They are referred to as trivial subgroups of G and all other subgroups of G are called non-trivial subgroups of G.

4.5.2 IDENTITIES AND INVERSES IN SUBGROUPS

Let G be a group with respect to the operation $*$ and let H be a subgroup of G. Then

(1)   the identity element of H is the identity of G.

(2)   the inverse of an element $h \in H$ is the inverse of h as an element of G.

4.5.3 CONDITIONS FOR SUBGROUPS (A)

Let G be a group with respect to the operation $*$, and let H be a subset of G. Then H is a subgroup of G if and only if

(i) H is not empty;

(ii)   if $a \in H$ and $b \in H$, then $a * b \in H$; and

(iii)   if $a \in H$, then $a^{-1} \in H$.

4.5.4 CONDITIONS FOR SUBGROUPS (B)

A subset H of a group G with respect to the operation $*$ is a subgroup of G if and only if

(i) H is not empty, and

(ii)   if $a \in H$ and $b \in H$, then $a * b^{-1} \in H$.

4.6   CYCLIC GROUPS

4.6.1   NOTATION

For elements a and b in a group G with respect to a binary operation $*$, it is the usual custom to write ab instead of $a * b$, and ab is read "a time b." We will follow this custom.

4.6.2   DEFINITION OF A CYCLIC GROUP

If G is a group with respect to the operation $*$, and there exists an element $a \in G$ such that

$$G = \{a^k \mid k \in Z\}$$

we say that G is a cyclic group and that G is generated by a or that a is a generator of G. This is denoted by $G = <a>$.

4.6.3   CYCLIC SUBGROUPS

A given subgroup K of a group G is called a cyclic subgroup of G if there exists an element $b \in G$ such that

$$K = <b> = \{y \in G \mid y = b^n \text{ where } n \in Z\}.$$

4.6.4   ORDER OF AN ELEMENT OF A GROUP

The order of an element a of a group G is the order of the cyclic subgroup generated by a. That is:

$$0(a) = 0(<a>).$$

Furthermore, an element $a \in G$ has order m if and only if m is the least positive integer such that $a^m = e$, where e is the identity element of G. If no such integer m exists, then a has infinite order.

4.6.5   FINITE CYCLIC GROUPS

If G is a cyclic group of finite order, say G has n elements, and a is a generator of G, then,

(i) $G = \{a^1, a^2, a^3, \ldots, a^n\}$, and $a^n = e$.

(ii)   every subgroup H of G is generated by an element of the form a^m where m is a divisor of n.

(iii)   the subgroup of G generated by a^r is also generated by a^d where $d = (r, n)$. In particular, the elements of G which generate G are those elements a^k with $(k, n) = 1$.

(iv)   if $a \in G$, then $0(a) = 0(a^{-1})$.

(v)   every subgroup H of G is itself a cyclic group.

4.7   GENERATING SUBSETS

If G is a group and $a \in G$, then the subgroup of G generated by a, $<a>$ is the smallest subgroup of G containing a.

4.7.1   THE INTERSECTION OF SUBGROUPS

Let C be any collection of subgroups of a group G; then the intersection of all the subgroups in C is a subgroup of G.

4.7.2   SMALLEST SUBGROUP CONTAINING A SUBSET

Let S be a subset of a group G, let C be the collection of all the subgroups of H in G such that $S \subseteq H$, and let D be the intersection of all the subgroups in C. Then D is a subgroup of G, and hence is the smallest subgroup of G.

4.7.3   SUBGROUPS GENERATED BY SUBSETS

Let G be a group with respect to an operation $*$ and let $S \subset G$. Then $<S>$ is the smallest subgroup of G such that $S \subset \, <S>$, and $<S>$ is called the subgroup of G generated by S.

4.8   GROUP HOMOMORPHISMS

Let G be a group with respect to the operation $*$, and let G' be a group with respect to the operation $\square$. A mapping $\theta : G \rightarrow G'$ is called a homomorphism from G to G' if

$$\theta(a * b) = \theta(a) \,\square\, \theta(b)$$

for all $a, b \in G$.

4.8.1   HOMOMORPHIC IMAGE

If there is a homomorphism from a group G onto a group G', then G' is said to be a homomorphic image of G.

4.8.2   EPIMORPHISMS AND MONOMORPHISMS

If θ is a homomorphism from a group G to a group G' that is surjective, then θ is called an epimorphism. If θ is injective, then θ is called a monomorphism.

4.8.3   ENDOMORPHISMS

A homomorphism from a group G to G itself is called an endomorphism.

4.8.4   KERNEL OF A GROUP HOMOMORPHISM

Let $\theta : G \rightarrow G'$ be a homomorphism from the group G (with respect to the operation $*$) to the group G' (with respect to the operation $\square$). Then the kernel ("ker") of θ is the set of all elements $x \in G$ such that $\theta(x) = e'$, where e' is the identity element of the group G'. That is, ker $\theta = \{x \in G \mid \theta(x) = e'\}$.

4.8.5   PROPERTIES OF GROUP HOMOMORPHISMS

Let G be a group with respect to the operation $*$ and identity element e; let G' be a group with respect to the operation $\square$ and identity element e'; and let $\theta : G \rightarrow G'$ be a homomorphism from G to G'. Then,

(1) $\theta(e) = e'$.

(2) $\theta(a^{-1}) = [\theta(a)]^{-1}$ for all $a \in G$.

(3) $\theta(a^n) = [\theta(a)]^n$ for all $a \in G$, and $n \in Z$.

(4) $\theta(G)$, the image of G under θ, is a subgroup of G'.

(5) $\theta(a) = \theta(b)$ if and only if $a * b^{-1} \in$ ker θ if and only if $a^{-1} * b \in$ ker θ.

(6) θ is injective implies ker $\theta = \{e\}$.

4.9   GROUP ISOMORPHISMS

Suppose that G is a group with respect to the operation $*$ and G' is a group with respect to the operation $\square$. A mapping $\alpha : G \rightarrow G'$ is called an isomorphism from G to G' if

(1) α is a bijection from G to G'. That is, α is one-to-one and onto.

(2) $\alpha(a * b) = \alpha(a) \ \square \ \alpha(b)$ for all $a, b \in G$. That is, α is a homomorphism from G to G'.

If an isomorphism from G to G' exists, we say that G is isomorphic to G'.

4.9.1   PROPERTIES OF GROUP ISOMORPHISMS

Let G be a group with respect to the operation $*$ and identity e; let G' be a group with respect to the operation $\square$ and identity e'; and let $\alpha : G \rightarrow G'$ be an isomorphism from G to G'. Then,

(a) $0(G) = 0(G')$.

(b)   if G is abelian, then G' is abelian.

(c)   if G is cyclic, then G' is cyclic.

(d)   if G has a subgroup of order n, then G' has a subgroup of order n.

(e)   if G has an element of order n, then G' has an element of order n.

(f)   if every element of G has its own inverse, then every element of G' has its own inverse.

(g)   if every element G is of finite order, then every element of G' is of finite order.

4.9.2   GROUP AUTOMORPHISM

A group isomorphism from a group G to itself is called a group automorphism of G.

4.10   PERMUTATION GROUPS

Let S be a non-empty set. A bijection (one-to-one and onto) mapping from S onto itself is called a permutation of S.

4.10.1   THE GROUP A(S) OF PERMUTATIONS OF S

Let S be any (finite or infinite) set, and let $A(S)$ consist of all the permutations of S. Then $A(S)$ is a group with respect to composition of mappings as the operation.

4.10.2   SUBGROUP OF AUTOMORPHISMS

Let $U(G)$ be the set of all automorphisms of a group G. Then $U(G)$ is a subgroup of the group of all permutations, $A(G)$, of G.

4.10.3   CONDITIONS FOR A(S) AND A(S') TO BE
ISOMORPHIC

Let S and S' be two non-empty sets, and let $\beta : S \rightarrow S'$ be a bijection from S to S'. Then for every permutation θ of S, the mapping $\beta^{-1} \theta \beta$ is a permutation of S'. Also, the mapping

$f: A(S) \rightarrow A(S')$ such that $f(\theta) = \beta^{-1} \theta \beta$ is an isomorphism from $A(S)$ to $A(S')$.

4.10.4   CALEY'S THEOREM

Let G be a (finite or infinite) group. Then the group $A(G)$ of all permutations of G has a subgroup K that is isomorphic to G.

Caley's theorem assures us that every group is isomorphic to a permutation group of its own elements. To find this group of permutations, for each $a \in G$, define a mapping $k_\alpha : G \rightarrow G$ by $k_\alpha(x) = ax$ for all $x \in G$. Then $K = \{k_a \mid a \in G$ is the permutation group isomorphic to the group G.

4.11   FINITE PERMUTATION GROUPS, S_n

Caley's theorem asserts that every group G is isomorphic to a subgroup of $A(G)$, the group of all permutations of the set G. If G is finite, and $0(G) = n$, then the group $A(G)$ is isomorphic to the group $A(\{1, 2, 3, ..., n\})$ which is called the symmetric group of degree n. This group is denoted by S_n.

4.11.1   ELEMENTS OF S_n

An element $\alpha \in S_n$ is a permutation on n symbols and is denoted by

$$\alpha = \begin{pmatrix} 1 & 2 & 3 & & n \\ \alpha(1) & \alpha(2) & \alpha(3) & \cdots & \alpha(n) \end{pmatrix}$$

That is, under each integer k, $1 \le k \le n$, put its image, $\alpha(k)$.

4.11.2   THE IDENTITY ELEMENT IN S_n

The permutation

$$e = \begin{pmatrix} 1 & 2 & 3 & \ldots & n \\ 1 & 2 & 3 & \ldots & n \end{pmatrix}$$

is called the identity permutation on S.

4.11.3   INVERSE OF A PERMUTATION

For any permutation

$$\alpha = \begin{pmatrix} 1 & 2 & \ldots & n \\ a_1 & a_2 & \ldots & a_n \end{pmatrix} \in S_n \, ,$$

the permutation

$$\beta = \begin{pmatrix} a_1 & a_2 & \ldots & a_n \\ 1 & 2 & \ldots & n \end{pmatrix} \in S_n$$

such that $\alpha\beta = e = \beta\alpha$, where e is the identity permutation on $S = \{1, 2, 3, \ldots, n\}$ is called the inverse of α and is denoted by α^{-1}.

4.11.4   PROPERTIES OF S_n

(1)   The order of the group S_n is $n!$ That is, $0(S_n) = n!$

(2)   If $n \geq 3$, then S_n is a non-abelian group.

(3)   The groups S_1 and S_2 are abelian.

(4)   If G is a finite group such that $0(G) = n$, then G is isomorphic to a subgroup of the group S_n.

4.11.5   CYCLE NOTATION, k-CYCLES

Let $S = \{1, 2, 3, ..., n\}$, let $A = \{a_1, a_2, ..., a_k\}$ be a subset of S, and let α be the permutation on S satisfying the following conditions:

(a) $\alpha(x) = x$ if $x \notin A$

(b) $\alpha(a_i) = a_{i+1}$ for $i = 1, 2, ..., k-1$

(c) $\alpha(a_k) = a_1$

Then α is called a k-cycle or a cycle of length k. This is represented by $(a_1, a_2, ..., a_k)$. Thus, if

$$\alpha = \begin{pmatrix} 1 & 2 & 3 & 4 & 5 \\ 1 & 4 & 2 & 3 & 5 \end{pmatrix} \in S_5 \, ,$$

then $\alpha = (432)$, which is a 3-cycle or a cycle of length 3.

4.11.6   TRANSPOSITIONS, 1-CYCLES

Every cycle of length 1 is the identity permutation, e, and any 2-cycle or a cycle of length 2 is called a transposition.

4.11.7   DISJOINT CYCLES

Two cycles $(a_1 a_2 ... a_r)$ and $(b_1 b_2 ... b_k)$ are said to be disjoint if they have no elements in common. Thus, cycles (123) and (456) are disjoint, whereas cycles (761) and (123) are not disjoint.

4.11.8   PRODUCT OF CYCLES

Cycles are usually composed like any other permutations except that the symbol "o" is omitted. Hence, a composition of cycles, or of other permutations, is referred to as a product.

4.11.9   PRODUCTS OF DISJOINT CYCLES

(1)   Every element $\theta \in S_n$ is either a cycle or a product of disjoint cycles.

(2)   Every element $\theta \in S_n$ is a product of transpositions. For example, if $\theta = (a_1\ a_2\ \ldots\ a_n)$, then $\theta = (a_1\ a_n)\ (a_2\ a_{n-1})\ \ldots (a_1\ a_2)$.

4.11.10   EVEN PERMUTATIONS, ODD PERMUTATIONS

A permutation $\alpha \in S_n$ is called an even permutation if α is a product of an even number of transpositions, and a permutation $\beta \in S_n$ is called an odd permutation if β is the product of an odd number of transpositions.

4.11.11   CONDITIONS FOR EVEN AND ODD PERMUTATIONS

An r-cycle, $(a_1\ a_2\ \ldots\ a_r)$, is an even permutation if r is odd and an odd permutation if r is even.

4.11.12   THE ALTERNATING GROUP, A_n

The subgroup of S_n that consists of all even permutations of S_n is called the alternating group, A_n.

The order of A_n is $\dfrac{n!}{2}$. That is, $0(A_n) = \dfrac{n!}{2}$.

CHAPTER 5

FURTHER TOPICS IN GROUP THEORY

5.1 COSETS OF A SUBGROUP

Let H be a subgroup of a group G, and let a be a fixed element of G. The set

$$aH = \{ah \mid h \in H\}$$

is called a left coset of H in G, and the set

$$Ha = \{ha \mid h \in H\}$$

is called a right coset of H in G.

If $+$ is the operation of G, then a left coset of H in G is written as

$$a + H = \{a + h \mid h \in H\},$$

and a right coset of H in G is written as

$$H + a = \{h + a \mid h \in H\}.$$

5.1.1 FACTS

If H is a subgroup of a group G, then each of the following is true:

(i)   If $Ha \cap Hb \neq \phi$, then $Ha = Hb$

(ii) $Ha = Hb$ if and only if $a \in Hb$.

5.1.2 EQUIVALENT CONDITIONS FOR RIGHT COSETS

If H is a subgroup of a group G, and $a, b \in G$, then the following conditions are equivalent:

(a) $ab^{-1} \in H$

(b) $a = hb$ for some $h \in H$

(c) $a \in Hb$

(d) $Ha = Hb$

As a consequence, the right coset of H to which a belongs is Ha.

5.1.3 EQUIVALENT CONDITIONS FOR LEFT COSETS

Let G be a group, let H be a subgroup of G, and let $a, b \in G$. Then the following conditions are equivalent:

(a) $a^{-1}b \in H$

(b) $a = bh$ for some $h \in H$

(c) $a \in bH$

(d) $aH = bH$

As a consequence, the left coset of H to which a belongs is aH.

5.1.4   DISTINCT COSETS OF A SUBGROUP

Let G be a finite group and H be a subgroup of G. To compute all the right (and similarly, all the left cosets) of H in G,

(1)   Write $H = He$

(2)   Select an element $a \in G$ such that $a \notin H$, and compute Ha

(3)   Select any element $b \in G$ such that $b \notin (H \cup Ha)$, and compute Hb

(4)   Continue in this way until all the elements of G are exhausted.

5.1.5   ORDER OF A COSET OF A SUBGROUP

If H is a finite subgroup of a group G, and $a \in G$, then

$$0(H) = 0(Ha), \text{ and } \quad 0(H) = 0(aH).$$

5.1.6   INDEX OF A SUBGROUP

Let G be a group of finite order and let H be a subgroup of G. Then the number of all distinct right (or left) cosets of H in G is called the index of H in G.

The index of H in G is denoted by $[G : H]$.

5.2 LAGRANGE'S THEOREM

Let G be a finite group of order n. If H is any subgroup of order m and index k, then $n = km$. In particular, both the order and index of any subgroup H of a finite group G are divisors of the order of G.

5.2.1 COROLLARIES

(1) If G is a finite group of order n, then the order of any element $a \in G$ divides n.

(2) A group G of prime order contains no subgroups other than $\{e\}$ and G itself.

(3) If G is a finite group of composite order, then G has non-trivial subgroups other than $\{e\}$ and G itself.

(4) Each group G of prime order is cyclic-generated by any one of its non-identity elements.

5.3 NORMAL SUBGROUPS

A subgroup N of a group G is said to be normal in G if every left coset of N in G is also a right coset of N in G. That is, $aN = Na$ for every element $a \in G$.

5.3.1 CONDITIONS FOR NORMALITY

A subgroup N of a group G is normal in G if and only if $g^{-1}ng \in N$ for every $n \in N$ and $g \in G$.

5.3.2   SUBGROUPS WITH INDEX 2

Let N be a subgroup of order m in a group G of order $2m$; that is, N has index 2 in the finite group G. Then N is a normal subgroup of G.

5.3.3   PRODUCT OF COSETS OF A NORMAL SUBGROUP

Let N be a normal subgroup in a group G. Then the product $aN \cdot bN$ of cosets aN and bN of N in G is the coset $(ab)N$.

It should be noted that if it is a subgroup of a group G, but H is not normal in G, then the product $aH \cdot bH$ of left cosets of H in G need not be a coset of H in G.

5.4   QUOTIENT GROUPS

Let N be a normal subgroup of a group G, and let G/N denote the set of all left cosets of N in G. Then G/N forms a group with respect to the operation $aN \cdot bN = (ab)N$ called the quotient group (or factor group) of G by N.

5.4.1   ORDER OF THE QUOTIENT GROUP G/N

Let N be a normal subgroup of a group G. Since the elements of the factor group G/N are the distinct left cosets of N in G, it follows that if G is a group of finite order, then the order of the group G/N is the index of N in G. Hence,

$$\text{order of } G/N = \frac{\text{order of } G}{\text{order of } N}.$$

THE NATURAL HOMOMORPHISM OF G ONTO *G/N*

Let N be a normal subgroup of a group G. Then the mapping $\alpha : G \rightarrow G/N$ defined by

$$\alpha(a) = aN$$

for each $a \in G$ is a homomorphism of G onto G/N with ker $\alpha = N$. The mapping α is called the natural homomorphism of G onto G/N.

5.5   DIRECT PRODUCTS OF GROUPS

If G_1, G_2 are two groups, then $G_1 \times G_2$ is the set of all ordered pairs (a, b) where $a \in G_1$ and $b \in G_2$ and where the product is defined by the components

$$(a_1, b_1)(a_2, b_2) = (a_1 a_2, b_1 b_2).$$

In general, if $G_1, G_2, \ldots, G_n$ are n groups, then their external direct product $G_1 \times G_2 \times \ldots \times G_n$ is the set of all n-tuples $(a_1, a_2, \ldots, a_n)$ where $a_i \in G_i$, $i = 1, 2, \ldots, n$, and where the product in $G_1 \times G_2 \times \ldots \times G_n$ is defined by the components

$$(a_1, a_2, \ldots, a_n)(b_1, b_2, \ldots, b_n) = (a_1 b_1, a_2 b_2, \ldots, a_n b_n).$$

5.5.1   INTERNAL DIRECT PRODUCT OF GROUPS

Let H and K be two normal subgroups of a group G. Then G is said to be the internal direct product of H and K, written $G = H \otimes K$ if and only if

(i) $G = HK$

(ii) $H \cap K = \{e\}$.

5.5.2   CONDITIONS FOR INTERNAL DIRECT PRODUCTS OF GROUPS

Let G be a group, and let H and K be normal subgroups of G. Then G is the internal direct product of its subgroups H and K if and only if

(1)   each element $x \in G$ can be uniquely expressed in the form $x = hk$, where $h \in H$ and $k \in K$

(2)   any element of H commutes with any element of K. That is, $hk = kh$ for each $h \in H$ and $k \in K$.

5.5.3   THE RELATIONSHIP BETWEEN INTERNAL DIRECT PRODUCT AND QUOTIENT GROUPS

If H and K are normal subgroups of the group G such that $G = H \otimes K$, then the quotient group G/H is isomorphic to K and the quotient group G/K is isomorphic to H.

5.5.4   GENERALIZATION OF INTERNAL PRODUCTS OF GROUPS

(1)   A group G is said to be the internal direct product of the normal subgroups $H_1, H_2, \ldots, H_n$ indicated by writing

$$G = H_1 \otimes H_2 \otimes \ldots \otimes H_n$$

        if

(i) $G = H_1 H_2 H_3 \ldots H_n$. That is, every element $x \in G$ is a product of the form

$$x = h_1 h_2 h_3 \ldots h_n$$

with $h_i \in H_i$, and

(ii) $H_i \cap H_i' = \{e\}$, where e is the identity element of G, and $H_i' = H_1 H_2 \ldots H_{i-1} H_{i-1} \ldots H_n$ for each i.

(2)   A group G is the internal direct product of the subgroups $H_1, H_2, \ldots, H_n$ if and only if

(i)   Every element $x \in G$ can be uniquely expressed in the form

$$x = h_1 h_2 \ldots h_n$$

with $h_i \in H_i$, $i = 1, 2, \ldots, n$.

(ii) $h_i \in H_i$ and $h_j \in H_j$, $i \neq J$, imply that $h_i h_j = h_j h_i$.

5.6   SEMI-DIRECT PRODUCTS OF GROUPS

Let N be a normal subgroup of a group G and let H be a subgroup of G. If $G = NH$ and $N \cap H = \{e\}$ where e is the identity element of G, then G is said to be a semi-direct product of N and H, and G splits over N, or is a split extension of N by H.

5.6.1   EXAMPLE OF A SEMI-DIRECT PRODUCT

Let R be the set of real numbers, and $R*$ be the set of nonzero

51

real numbers. Then the group $G = R \times R^*$ with multiplication given by

$$(a, b)(c, d) = (a + bc, bd)$$

is a split extension of the normal subgroup $N = \{(a, 1) \mid a \in R\}$ by the subgroup $H = \{(0, b) \mid b \in R^*\}$.

5.6.2   THEOREMS (SEMI-DIRECT PRODUCTS)

Let N and H be groups and let θ be a homomorphism from H to the automorphism group $\text{Aut}(N)$ of N. Then

(a)   The product $G = N \times H$ is a group under the operation

$$(n_1, h_1)(n_2, h_2) = (n_1 \, \theta \, (h_1) \, n_2, h_1 h_2)$$

(b) $N^* = \{(n, e) \mid n \in N\}$ is a normal subgroup, and $H^* = \{(e, h) \mid h \in H\}$ is a subgroup of G

(c) G is a split extension of its normal subgroup N^* by its subgroup H^*.

5.7   FINITE ABELIAN GROUPS

When dealing with abelian groups, it is customary to use addition notation. That is, the group operation is called "addition," and we "add" two elements rather than "multiply" them. We write $g + h$ instead of $g \cdot h$ or gh and ng replaces g^n. Also, direct products are called direct sums. Thus, if $G_1, G_2, \ldots, G_n$ are n abelian groups, then their direct sum is written as

$$G_1 \oplus G_2 \oplus \ldots \oplus G_n.$$

5.7.1   SUMS OF SUBGROUPS OF ABELIAN GROUPS

If $H_1, H_2, \ldots, H_n$ are subgroups of an abelian group G, then the sum $H_1 + H_2 + \ldots + H_n$ of these subgroups is defined by

$$H_1 + H_2 + \ldots + H_n = \{x \in G \mid x = h_1 + h_2 + \ldots + h_n$$
$$\text{with } h_i \in H_i\}$$

and is a subgroup of G.

5.7.2   DIRECT SUMS OF SUBGROUPS OF ABELIAN GROUPS

Let $H_1, H_2, \ldots, H_n$ be n subgroups of an abelian group G. Then G is called the direct sum of the subgroups $H_1, H_2, \ldots, H_n$ written

$$G = H_1 \oplus H_2 \oplus H_3 \oplus \ldots \oplus H_n$$

if and only if every element $g \in G$ can be expressed uniquely in the form

$$g = h_1 + h_2 + \ldots, h_n$$

where $h_i \in H_i$, $i = 1, 2, 3, \ldots, n$.

5.7.3   ORDER OF A DIRECT SUM OF SUBGROUPS

If $H_1, H_2, \ldots, H_n$ are finite subgroups of the abelian group G such that their sum is direct, then the order of $H_1 \oplus H_2 \oplus \ldots \oplus H_n$ is the product of the orders of the subgroups H_i. That is,

$$0(H_1 \oplus H_2 \oplus \ldots \oplus H_n) = 0(H_1)0(H_2) \ldots 0(H_n).$$

5.7.4   *p*-GROUPS

If p is a prime, then a group G is called a p-group if each of its elements has order that is a power of p.

5.7.5   THE SUBGROUP G_p

If G is a finite abelian group that has order that is divisible by a prime p, then the set G_p is defined as the set of all elements $x \in G$ that have orders that are powers of p. This subset G_p of G is a subgroup of G.

5.7.6   FINITELY GENERATED ABELIAN GROUPS

An abelian group G is said to be finitely generated if there exists a set of elements $\{a_1, a_2, \ldots, a_n\}$ in G such that every $x \in G$ can be written in the form

$$x = z_1 a_1 + z_2 a_2 + \ldots + z_n a_n$$

where the zi's are all integers. The elements $a_1, a_2, \ldots, a_n$ are called generators of G and the set $\{a_1, a_2, \ldots, a_n\}$ is called a generating set for G.

5.7.7   FUNDAMENTAL THEOREM ON FINITE GROUPS

Any finitely-generated abelian group G (and therefore any finite abelian group) is a direct sum of cyclic groups.

5.8   SYLOW *p*-SUBGROUPS

If p is a prime and m is a positive integer such that $p^m \nmid 0(G)$

and $p^{m+1} \nmid 0(G)$, then a subgroup of G that has order p^m is called a Sylow p-subgroup.

5.8.1   CAUCHY'S THEOREM FOR ABELIAN GROUPS

If G is an abelian group of order n and p is a prime such that $p|n$, then G has at least one element of order p.

5.8.2   THE SYLOW THEOREMS

Let G be a finite group and let p be a prime integer.

(1)   If m if a positive integer such that $p^m \mid 0(G)$ and $p^{m+1} \nmid 0(G)$, then G has a subgroup of order p^m.

(2)   If H and K are any two Sylow p-subgroups for the same prime p, then there is an element $a \in G$ such that

$$K = \{a^{-1}ha \mid h \in H\}, H = \{aka^{-1} \mid k \in K\}.$$

That is, H and K are conjugate subgroups.

(3)   If $p|0(G)$, then the number of Sylow p-subgroups of G, n_p, is an integral divisor d of the $0(G)$ satisfying the property

$$d \equiv 1 \pmod{p}.$$

That is, $d = 1 + k_p$, where k is a non-negative integer.

CHAPTER 6

RING THEORY

6.1   DEFINITION OF A RING (A)

A ring is a nonempty set R on which there are defined two binary operations called "addition" and "multiplication," and denoted by + and · respectively, such that

(1) R is closed under addition: $a, b \in R$ implies $a + b \in R$.

(2)   Addition in R is associative: for all $a, b, c \in R$, $(a + b) + c = a + (b + c)$.

(3) R has an additive identity: there exists an element $0 \in R$ such that $a + 0 = 0 + a = a$ for all $a \in R$.

(4) R has additive inverses: given $a \in R$, there exists an element $b \in R$ such that $a + b = 0$. The element $b \in R$ is called the additive inverse of a and written as $-a$.

(5)   Addition in R is commutative: $a + b = b + a$ for all $a, b \in R$.

(6) R is closed under multiplication: $a, b \in R$ implies $a \cdot b$ $\in R$.

(7)   Multiplication in R is associative: for all $a, b, c \in R$, $(a \cdot b) \cdot c = a \cdot (b \cdot c)$.

(8)   Two distributive laws hold in R:

$$a \cdot (b + c) = a \cdot b + a \cdot c$$

and

$$(b + c) \cdot a = b \cdot a + c \cdot a$$

for all $a, b, c \in R$.

6.1.1   DEFINITION OF A RING (B)

A ring is a nonempty set R on which there are defined two operations called "addition" and "multiplication," and denoted by $+$ and $\cdot$ respectively, such that the following three axioms are satisfied:

A_1) R is an abelian group with respect to addition, $+$.

A_2) R is closed and associative under multiplication.

A_3)   Multiplication is distributive over addition. That is, for all $a, b, c \in R$,

$a \cdot (b + c) = a \cdot b + a \cdot c$ and $(b + c) \cdot a = b \cdot a + c \cdot a$

6.1.2   EXAMPLES OF RINGS

(1)   Some simple examples of rings are provided by the familiar number systems with their usual operations of addition and multiplication. These are:

   (a)   The set Z of all integers.

   (b)   The set Q of all rational numbers.

   (c)   The set R of all real numbers.

   (d)   The set C of all complex numbers.

(2)   The set E of all even integers is a ring with respect to the usual addition and multiplication in Z.

(3)   The set of all 2×2 matrices over the set of all real numbers is a ring with respect to addition and multiplication of matrices of real numbers.

(4)   Let $R = \{u, v, w, x\}$. Define addition and multiplication in R by means of the following tables:

+	u	v	w	x
u	u	v	w	x
v	v	u	x	w
w	w	x	u	v
x	x	w	v	u

$\cdot$	u	v	w	x
u	u	u	u	u
v	u	v	w	x
w	u	w	w	u
x	u	x	u	x

R is a ring with respect to these two operations "+" and "·".

6.1.3   COMMUTATIVE RINGS

A commutative ring is a ring R in which multiplication is a commutative operation; that is, for all $a, b \in R, a \cdot b = b \cdot a$. In case $a \cdot b = b \cdot a$ for a particular pair of elements $a, b \in R$, we express this by saying that a and b commute.

6.1.4   RINGS WITH IDENTITY

A ring with identity is a ring R in which there exists an identity element (also called a unity) for the operation multiplication, normally represented by the symbol e so that $a \cdot e = e \cdot a = a$ for all $a \in R$.

6.1.5   INVERTIBLE ELEMENTS OF A RING WITH UNITY

Let R be a ring with unity e. An element $x \in R$ for which there exists an element $y \in R$ such that $xy = e = yx$ is called an invertible or a unit element of R. The multiplicative inverse y of x is also denoted by x^{-1} and is called the reciprocal of x.

6.1.6   GENERALIZED ASSOCIATIVE LAWS

Let $n \geq 2$ be a positive integer and let $a_1, a_2, \ldots, a_n$ denote elements of a ring R. Then for any positive integer m such that $1 \leq m < n$,

(i) $(a_1 + a_2 + \ldots + a_m) + (a_{m+1} + a_{m+2} + \ldots + a_n)$
         $= a_1 + a_2 + \ldots + a_{m+1} + \ldots + a_n$

(ii) $(a_1 a_2 \ldots a_m)(a_{m+1} a_{m+2} \ldots a_n)$
         $= a_1 a_2 \ldots a_m a_{m+1} \ldots a_n.$

GENERALIZED DISTRIBUTIVE LAWS

Let $n \geq 2$ be a positive integer and let $b, a_1, a_2, \ldots, a_n$ denote elements of a ring R. Then

(i) $\quad b(a_1 + a_2 + \ldots + a_n) = ba_1 + ba_2 + \ldots + ba_n$

(ii) $\quad (a_1 + a_2 + \ldots + a_n)b = a_1 b + a_2 b + \ldots + a_n b.$

6.2 PROPERTIES OF RINGS

Let R be a ring, 0 the zero element of R, e the identity of R, and $a, b, c \in R$. Then each of the following is true:

(1) The element 0 of R is unique.

(2) Each element of R has a unique additive inverse.

(3) If $a + b = a + c$, then $b = c$ (Left Cancellation Law).

(4) If $b + a = c + a$, then $b = c$ (Right Cancellation Law).

(5) Each of the equations $a + x = b$ and $x + a = b$ has a unique solution.

(6) $-(-a) = a$ and $-(a + b) = (-a) + (-b)$ for all $a, b \in R$.

(7) If m and n are integers, then

$$(m + n)a = ma + na, \, m(a + b) = ma + mb, \text{ and}$$
$$m(na) = (mn)\, a.$$

(8) $0 \cdot a = a \cdot 0 = 0$.

(9) $a(-b) = (-a)b = -(ab)$.

(10) $(-e)a = -a$.

(11) $(-a)(-b) = ab$.

(12) $a(b-c) = a(b+(-c)) = ab + (-ac) = ab - ac$, and $(a-b)c$
        $= ac - bc$.

(13)   If an element $a \in R$ has a multiplicative inverse, it is
        unique.

(14)   The unity element, e, of R is unique.

6.3   SUBRINGS

A nonempty set S of a ring R is a subring of R if S is itself a ring
with respect to the operations on R.

6.3.1   SUBRING CONDITIONS (A)

Let R be a ring and S be a nonempty subset of R. Then S is a
subring of R if and only if the following conditions hold:

(a) $x \in S$ and $y \in S$ imply $x + y \in S$ and $xy \in S$.

(b) $x \in S$ implies $-x \in S$.

6.3.2   SUBRING CONDITIONS (B)

A subset S of a ring R is a subring of R if and only if these conditions hold:

(a) S is not empty.

(b) $x \in S, y \in S$ imply $x - y \in S$ and $xy \in S$.

6.4   INTEGRAL DOMAINS

6.4.1   ŻERO-DIVISORS AND REGULAR ELEMENTS IN A RING

An element $a \neq 0$ in a commutative ring R is called a zero-divisor in R if there exists an element $b \neq 0$ in R such that $ab = 0$.

An element of R that is not a zero-divisor is called a regular element.

6.4.2   CANCELLATION LAW OF MULTIPLICATION

If an element $a \in R$ is not a divisor of zero in R, then each of the following holds:

(a)   If $b, c \in R$ such that $ab = ac$, then $b = c$.

(b)   If $b, c \in R$ such that $ba = ca$, then $b = c$.

6.4.3   INTEGRAL DOMAIN

A ring D with a unity element $e \neq 0$ is called an integral domain if D is commutative, and has no divisors of zero.

6.4.4   ADDITIVE ORDER OF ELEMENTS OF AN INTEGRAL DOMAIN

In the additive group of an integral domain D, either all the elements except zero have infinite order or all the nonzero elements have the same finite order.

6.4.5   CONDITION FOR AN INTEGRAL DOMAIN

Let D be an integral domain with unity $e \neq 0$ in which each of $ac = bc$ and $ca = cb$ implies that $c = 0$ or $a = b$. Then D is an integral domain.

6.5   DIVISION RINGS

A ring R with unity e is said to be a division ring if for each element $a \in R$, there exists an element $b \in R$ (usually written as a^{-1}) such that

$$aa^{-1} = a^{-1}a = e.$$

6.5.1   CONDITIONS FOR DIVISION RINGS

(a)   A ring R is a division ring if the set R^* of all nonzero elements of R is a group with respect to the operation multiplication.

(b)   A finite ring R with unity element, e, and no divisors of zero is a division ring.

6.6 FIELDS

Let F be a ring. Then F is a field provided the following conditions hold:

(1) F is a commutative ring.

(2) F has unity element e such that $e \neq 0$.

(3)   Every nonzero element of F has a multiplicative inverse.

Equivalently, a field is a commutative division ring.

6.6.1   ALTERNATE CHARACTERIZATION OF A FIELD

A field is a set of elements on which there are defined two operations called "addition" and "multiplication" and denoted by $+$ and $\cdot$ respectively, such that the following conditions hold:

(i) F forms an abelian group with respect to addition.

(ii)   The nonzero elements of F form an abelian group with respect to multiplication.

(iii)   The distributive law holds; that is,

$$x(y + z) = xy + xz$$

for all $x, y, z \in F$.

6.6.2   THE RELATION BETWEEN FIELDS AND INTEGRAL DOMAINS

(a)   Every field is an integral domain.

(b)   Every finite integral domain is a field.

6.6.3   SUBFIELDS, EXTENSION FIELDS

A subset F of a field E is a subfield in E if F itself is a field with respect to the operations of F. If F is a subfield of a field E, then E is said to be an extension field, or superfield over F.

6.6.4   SUBFIELD CONDITIONS

Let F be a subfield of at least two elements in a field E. Then F is a subfield of E if and only if F is closed with respect to subtraction and division by nonzero elements. That is,

(i) $a - b \in F$ for all $a, b \in F$

(ii) $a \div b \in F$ for all $a, b \in F$, and $b \neq 0$.

6.7   ORDERED INTEGRAL DOMAINS

An integral domain D is said to be an ordered integral domain if D contains a subset D^+ with the following properties:

(1) D^+ is closed under addition: if $a, b \in D^+$, then $a + b \in D^+$

(2) D^+ is closed under multiplication: if $a, b \in D^+$, then $a \cdot b \in D^+$

(3)   For each element $a \in D$, one and only one of the following holds:

$$a = 0, \quad a \in D^+, \quad -a \in D^+.$$

6.7.1 POSITIVE ELEMENTS, NEGATIVE ELEMENTS

Let D be an integral domain by a given subset D^+. For this ordering, the elements of D^+ are called the positive elements of D. If $a \in D$ such that $-a \in D^+$, then a is called a negative element of D.

6.7.2 SOME PROPERTIES OF ORDERED INTEGRAL DOMAINS

Let D be an ordered integral domain with D^+ as the set of positive elements of D. Then

(1)   If $a \in D$ and $a \neq 0$, then $a^2 \in D^+$

(2)   The unity element e of D is an element of D^+

(3)   If $a \in D^+$ and n is a positive integer, then $na \in D^+$

(4) D contains a subring isomorphic to the ring of all integers, Z.

6.7.3 LEAST ELEMENT OF A SUBSET OF AN ORDERED INTEGRAL DOMAIN

Let D be an ordered integral domain. An element m in a subset S of D is called a least element of S if $m < x$ for each element $x \in S$ such that $x \neq m$.

6.7.4 WELL-ORDERED INTEGRAL DOMAINS

An ordered integral domain D with D^+ as the set of positive elements is said to be well-ordered if every nonempty set of D^+ has a least element.

6.7.5 CHARACTERIZATION OF THE RING OF INTEGERS

Let D be an ordered integral domain in which the set D^+ of positive elements is well-ordered. Then

(i)   the unity element of D is the least element of D^+.

(ii) $D^+ = \{ ne \mid n \in Z^+ \}$.

(iii) D is isomorphic to the ring of integers, Z.

6.8   IDEALS

Let I be a subring of a ring R. Then

(i) I is said to be a right ideal in R if I is closed with respect to multiplication on the right by elements of R. That is, if $a \in I$ and $r \in R$, then $ar \in I$.

(ii) I is said to be a left ideal in R if I is closed under multiplication on the left by elements of R. That is, if $a \in I$ and $r \in R$, then $ra \in I$.

(iii) I is said to be an ideal in R if I is both a right ideal and a left ideal in R. That is, it is closed with respect to multiplication on either side by elements of R. If $a \in I$ and $r \in R$, then $ar \in I$ and $ra \in I$.

6.8.1   PRINCIPAL IDEAL

Let R be a commutative ring with unity and let a be a fixed element of R. Then, the ideal

$$(a) = \{ar \mid r \in R\}$$

which consists of all multiples of a by elements r of R is called the principal ideal generated by a in R.

6.8.2 IDEALS IN THE RING OF INTEGERS, Z

In the ring, Z, of integers, every ideal is a principal ideal.

6.8.3 COSETS OF AN IDEAL

Let I be an ideal in a ring R and let a be any element of R. Since R is an abelian group under addition, it follows that I is a normal subgroup of this addition group. Hence, the ideal coset for the element a of the ideal I in the ring R is the same subset of R as the coset $a + I$ considering I to be a normal subgroup in the additive group of R.

6.9 QUOTIENT RINGS

6.9.1 THE RING OF COSETS OF AN IDEAL

Let I be an ideal of the ring R. Then the set R/I of all additive cosets $r + I$ of I in R forms a ring with respect to coset addition:

$$(a + I) + (b + I) = (a + b) + I$$

and coset multiplication:

$$(a + I)(b + I) = (a\,b) + I.$$

6.9.2 QUOTIENT RING

If I is an ideal of the ring R, then the ring R/I described in 6.9.1 above is called the quotient ring of R by I.

6.9.3   IDEALS OF A FIELD

If F is a field, then F has no ideals other than (0) and F itself.

6.10   RING HOMOMORPHISM

A mapping ϕ from a ring R to a ring R' such that for all $a, b \in R$,

(a) $\phi(a + b) = \phi(a) + \phi(b)$

(b) $\phi(a\,b) = \phi(a)\,\phi(b)$,

is called a homomorphism from R to R'.

6.10.1   EPIMORPHISM, HOMOMORPHIC IMAGE

Let ϕ be a homomorphism from a ring R to a ring R'. If ϕ is surjective, then ϕ is called an epimorphism and R' is called a homomorphic image of R.

6.10.2   RING ISOMORPHISM

A bijective homomorphism from a ring R to a ring R' is called a ring isomorphism.

6.10.3   ENDOMORPHISMS, AUTOMORPHISMS

(i)   A ring homomorphism from a ring R to R itself is called a ring endomorphism of R.

(ii)   A ring isomorphism from a ring R to R itself is called a ring automorphism.

6.10.4 KERNEL OF A RING HOMOMORPHISM

Let ϕ be a homomorphism from a ring R to a ring R'. Then the set

$$\ker \phi = \{r \in R \mid \phi(r) = 0_{R'}\}$$

where $0_{R'}$ being the zero element of R', is called the kernel of ϕ.

6.10.5 PROPERTIES OF RING HOMOMORPHISM

Let ϕ be a homomorphism from a ring R to the ring R'. Then

(1) $\phi(0_R) = 0_{R'}$

(2) $\phi(-r) = -\phi(r)$ for all $r \in R$

(3) If S is a subring of R, then $\phi(S)$ is a subring of R'

(4) If S' is a subring of R', then $\phi^{-1}(S')$ is a subring of R

(5) Ker ϕ is an ideal of R

(6) Ker $\phi = \{0_R\}$ if and only if ϕ is injective.

6.11 RELATIONSHIP BETWEEN QUOTIENT RINGS, RING HOMOMORPHISMS AND IDEALS

6.11.1 NATURAL RING HOMOMORPHISM

If R is a ring and I is an ideal in R, then the mapping

$$\alpha : R \to R/I$$

defined by

$$\alpha(r) = r + I$$

for each $r \in R$ is called the natural homomorphism from R onto R/I and ker $\alpha = I$.

6.11.2 FUNDAMENTAL HOMOMORPHISM THEOREM FOR RINGS (FIRST HOMOMORPHISM THEOREM)

Let R and R' be rings, and let $\alpha : R \to R'$ be a homomorphism from R onto R' with ker $\alpha = I$. Then the mapping

$$\phi : R/I \to R'$$

defined by

$$\phi(r + I) = \alpha(r)$$

for each $r + I \in R/I$ is an isomorphism of R/I onto R'. Therefore, R/I is isomorphic to R'.

6.11.3 SECOND HOMOMORPHISM THEOREM

Let the mapping $\phi : R \to R'$ be a homomorphism from a ring R onto a ring R' with kernel K. If I' is an ideal in R', let $I = \{a \in R \mid \phi(a) \in I'\}$. Then I is an ideal of R, $I \supset K$, and I/K is isomorphic to I'. This sets up a one-to-one correspondence between all the ideals of R' and those ideals of R that contain K.

6.11.4 THIRD HOMOMORPHISM THEOREM

Let the mapping $\phi : R \to R'$ be a homomorphism from R onto R' with kernel K. If I' is an ideal of R' and $I = \{a \in R \mid \phi(a) \in I'\}$, then R/I is isomorphic to R'/I'. Equivalently, if K is an ideal of R and $I \supset K$ is an ideal of R, then R/I is isomorphic to $(R/I)/(I/K)$.

6.12 CHARACTERISTIC OF A RING

Let R be a ring. If there exists a positive integer m such that $ma = 0$ for every element $a \in R$, then the smallest such positive integer m is called the characteristic of R. If no such positive integer m exists, then R is said to have characteristic zero.

6.12.1 CHARACTERISTIC OF A RING WITH UNITY, *e*

Let R be a ring with unity e. If there exists a positive integer m such that $me = 0$, then the smallest such positive integer m is the characteristic of R. If no such positive integer m exists, then R has characteristic zero.

6.12.2 CHARACTERISTIC OF AN INTEGRAL DOMAIN

If D is an integral domain, then the characteristic of D is either zero or a prime number.

6.12.3 EMBEDDING A RING IN AN INTEGRAL DOMAIN

Let D be an integral domain.

(a)   If D is of characteristic zero, then D contains a subring isomorphic to the ring of all integers, Z

(b)   If D is of characteristic p, where p is a prime, then D contains a subring isomorphic to Z_p.

That is, any integral domain of characteristic zero has Z embedded in it, and any integral domain with prime characteristic has Z_p embedded in it.

6.13   MAXIMAL IDEALS

A maximal ideal in a ring R is an ideal M of R such that $M \neq R$, and whenever I is an ideal of R such that $M \subset I \subset R$, then $I = M$ or $I = R$.

6.13.1   PRIME IDEALS

An ideal I of a ring R is said to be a prime ideal if, for all a, b $\in R$, $ab \in I$ implies $a \in I$ or $b \in I$.

6.13.2   IDEALS IN COMMUTATIVE RINGS

Let R be a commutative ring with unity, e, and let M be an ideal of R. Then

(1)   the quotient ring R/M is a field if and only if M is a maximal ideal of R

(2)   every maximal ideal I of R is a prime ideal.

THE PROBLEM SOLVERS

The "PROBLEM SOLVERS" are comprehensive supplemental textbooks signed to save time in finding solutions to problems. Each "PROBLEM SOLVER" is the of its kind ever produced in its field. It is the product of a massive effort to illustrate al any imaginable problem in exceptional depth, detail, and clarity. Each problem is wo out in detail with step-by-step solution, and the problems are arranged in order of comple from elementary to advanced. Each book is fully indexed for locating problems rapid

ADVANCED CALCULUS
ALGEBRA & TRIGONOMETRY
AUTOMATIC CONTROL
   SYSTEMS/ROBOTICS
BIOLOGY
BUSINESS, MANAGEMENT,
   & FINANCE
CALCULUS
CHEMISTRY
COMPLEX VARIABLES
COMPUTER SCIENCE
DIFFERENTIAL EQUATIONS
ECONOMICS
ELECTRICAL MACHINES
ELECTRIC CIRCUITS
ELECTROMAGNETICS
ELECTRONIC COMMUNICATIONS
ELECTRONICS
FINITE & DISCRETE MATH
FLUID MECHANICS/DYNAMICS
GENETICS

GEOMETRY:
PLANE • SOLID • ANALYTIC
HEAT TRANSFER
LINEAR ALGEBRA
MACHINE DESIGN
MECHANICS : STATICS • DYNAMICS
NUMERICAL ANALYSIS
OPERATIONS RESEARCH
OPTICS
ORGANIC CHEMISTRY
PHYSICAL CHEMISTRY
PHYSICS
PRE-CALCULUS
PSYCHOLOGY
STATISTICS
STRENGTH OF MATERIALS &
   MECHANICS OF SOLIDS
TECHNICAL DESIGN GRAPHIC:
THERMODYNAMICS
TRANSPORT PHENOMENA :
MOMENTUM • ENERGY • MASS
VECTOR ANALYSIS

If you would like more information about any of these books, complete the cou below and return it to us or go to your local bookstore.

RESEARCH and EDUCATION ASSOCIATION
61 Ethel Road W. • Piscataway • New Jersey 08854
Phone: (201) 819-8880

Please send me more information about your Problem Solver Books

Name _____

Address _____

City _____ State _____ Zip _____